YOUR KNOWLEDGE HAS VALUE

- We will publish your bachelor's and master's thesis, essays and papers

- Your own eBook and book - sold worldwide in all relevant shops

- Earn money with each sale

Upload your text at www.GRIN.com
and publish for free

Bibliographic information published by the German National Library:

The German National Library lists this publication in the National Bibliography; detailed bibliographic data are available on the Internet at http://dnb.dnb.de .

This book is copyright material and must not be copied, reproduced, transferred, distributed, leased, licensed or publicly performed or used in any way except as specifically permitted in writing by the publishers, as allowed under the terms and conditions under which it was purchased or as strictly permitted by applicable copyright law. Any unauthorized distribution or use of this text may be a direct infringement of the author s and publisher s rights and those responsible may be liable in law accordingly.

Imprint:

Copyright © 2018 GRIN Verlag
Print and binding: Books on Demand GmbH, Norderstedt Germany
ISBN: 9783668725720

This book at GRIN:

https://www.grin.com/document/428652

Leonard Kahungu

Strategic Decisions in Managing Energy Resources

GRIN Verlag

GRIN - Your knowledge has value

Since its foundation in 1998, GRIN has specialized in publishing academic texts by students, college teachers and other academics as e-book and printed book. The website www.grin.com is an ideal platform for presenting term papers, final papers, scientific essays, dissertations and specialist books.

Visit us on the internet:

http://www.grin.com/

http://www.facebook.com/grincom

http://www.twitter.com/grin_com

STRATEGIC DECISIONS IN MANAGING ENERGY RESOURCES

External Factors Influencing GPRE's Strategic Decision 2
   Political Factors 2
   Economic Factors... 3
   Social Factors... 3
   Technological Factors 4
   Legal Factors.. 4
   Environmental Factors 5
GPRE's External Competitive Advantages 5
   Threats to New Entrants.. 5
   Threats of Rivals 5
   Threats of Bargaining Power of Buyers.. 6
   Threats of Bargaining power of Suppliers 6
   Threats of Substitutes.. 6
GPRE's Competitive Strengths.. 6
Conclusion 7
References.. 8

There have been major concerns regarding the efficiency in relying on fossil fuels in the production and management of energy in the contemporary world. These concerns are attributed to heated debates regarding the role of fossil fuels in global warming, environmental population, and sustainability. Most organisations in the energy sector are increasingly focusing on clean energy and efficient management of energy resources to achieve secure and sustainable production of power, with some transforming from relying on fossil-powered stations to relying on renewable power sources (Prpich, Darabkhani, Oakey, & Pollard, 2014). In particular, the Green Plains Inc. in Omaha, Nebraska, USA, is one of the classic examples of energy organisations shifting from depending on oil and gas sector to exclusive focus on renewable energy production due to external and competitive factors influencing the company's strategic business decisions. The Green Plains Inc., commonly abbreviated as GPRE in the stocks market, has emerged as one of the leading biofuel producer in the North American region (Mathiyazhagan & Ganapathi, 2011). There exists a list of several justifications indicating GPRE's strategic decision to cease from relying on fossil fuels into using, producing, and distributing renewable energy resources.

External Factors Influencing GPRE's Strategic Decision

Political Factors

At the national level, the U.S Department of Energy (DoE) plays a central role in influencing the formulation of energy regulations that predominantly focus on energy conservation. Thus, the US DoE affects political decisions affecting the energy management in attempts to achieve a high level of conservation. The state governments are also empowered by the Renewable Portfolio Standards (RPS) to encourage the adoption of renewable energy sources. As a consequence, the Nebraska Oil and Gas Conservation Commission along with other energy regulations in the state provides funding and financial benefits to enhance investment in renewable energy resources.

Nebraska Energy Office administers performance-based benefits, tax incentives, grants, and rebate programs among other RPS programs to organisations dealing with renewable sources of energy. The US is also obliged by international and regional bodies to facilitate the enactment of policies aimed at accomplishing clean air obligations by supporting research and investment in renewable energy sources (Becker, 2016). It is no coincidence that GPRE is located in Nebraska since the state reduces various barriers attributed to investment risks, compliance charges and regulations affecting the adoption or production of renewable energy for residential and industrial purposes. Therefore, political stability and favourable political factors in at the federal, state, and local levels have contributed to GPRE strategic decision to rely and produce on biofuels as opposed to fossil fuel.

Economic Factors

Economic factors affecting the energy industry include government expenditure, labour policies, inflation, stages of the business cycle, monetary policies and the consumer confidence. The current economic stability in the US is relatively stable, which favours the development of emerging businesses. Nevertheless, demand for biofuels has significantly increased in the recent past increased within the US market and across the world. According to one of the latest GPRE's annual report released in 2016 the demand and supply for biofuels, particularly ethanol has seen a significant increase on demand and supply (Liserre, Sauter, & Hung, 2015). Favourable federal government programs such as the United States Department of Agriculture acreage regulations and prices have played a central role in supporting GPRE's activities. Furthermore, the US federal government mandates the production and use of renewable fuels through the Renewable Fuel Standards, which has significantly increased domestic production of biofuels. The US has also been identified as one of the leading ethanol exporters to the global market due to favourable multilateral economic agreements (Kehbila, 2012). The presence of the outlined economic factors has been central to the GPRE's expansion and growth in renewable energy sources.

Social Factors

The increasing demand for renewable energy products indicates changing attitude patterns in the contemporary society. Besides, the debate concerning the role of fossil fuel in environmental population has gradually altered social perceptions regarding energy resources in favour of renewable sources (Edomah, Foulds, & Jones, 2016). Highly volatile market prices have also contributed to common attitudes towards biofuels. As a consequence, these changes in the society are some of the key drivers in the consumption of biofuels. This indicates that social developments may have significantly influenced GPRE's goals in the production of renewable energy sources (Liserre, Sauter, & Hung, 2015). Population growth across the globe may also provide a large market pool for renewable energy sources.

Technological Factors

In the wake of declining fossil fuel reserves and increased concerns regarding environmental conservation, industries are progressively focusing on technological advancements in attempts to develop products relying on adaptive energy demands. For instance, the production of automotive relying on both gasoline and biofuels is essential in fuelling renewable energy products (Goss, Millewski, & Strain, 2015). This may have resulted in an increased demand for biofuels and hence influencing the production of renewable energy (Lantz, 2016). Additionally, GPRE has deployed vast resources in research and development activities in attempts to develop advanced products resulting in a higher efficiency output compared to ethanol production (Becker, 2016). Successful developments may lead to highly effective biofuel products, which may increase the market shares as the consumers strive in searching highly efficient and reliable energy sources.

Legal Factors

The federal, state and local regulations provide a legislative framework affecting the energy industry. These regulations influence the nature and scope of investment in sustainable energy sources. International regulations affect the rate of adopting renewable energy utilities. Presently, the legal framework in the United States requires companies dealing with renewable energy sources to comply with minimal standards before accessing predetermined benefits accorded by the local, state, and the federal government as stipulated by the Renewable Fuel Standards (Liserre, Sauter, & Hung, 2015). These factors have compelled organisations in renewable sources energy into exploring their capacity limits in efforts to access tax and monetary incentives. Besides, the US has established bilateral trade agreements that allow the exportation of biofuels into other regions, particularly the European Union region. International regulations affecting the energy industry have increasingly become vocal regarding the increment of alternative energy sources. This has resulted in an expansion of the market demand for biofuels. These benefits at the state, federal and international level contain lucrative advantages, encouraging high-risk ventures into biofuels (Kehbila, 2012). This may have encouraged the Green Plains Inc. to expand its biofuel production capacity, making it one of the largest producer and exporter of biofuels in the North American and Eurasia regions.

Environmental Factors

Fossil fuels are associated with hazardous emissions that affect the environment. These emissions are thought to be the leading contributors to global warming. As a consequence, international organisations and other key players in the energy sector have become vocal in encouraging clean energy utilities. This has resulted in the enactment of policies seeking to conserve the environment while encouraging the production of alternative sources of energy, other than the fossil fuel products (Maxson et al., 2017). Environmental policies encouraging renewable source provide economic, research, political, and market incentives. These factors may have encouraged GPRE to invest in renewable sources of energy in attempts to tap the untapped marketing potential nurtured by environmental concerns.

GPRE's External Competitive Advantages

Threats to New Entrants

The industrial energy segment dealing with renewable sources of energy is highly fragmented and lacks an explicit pattern among the new entrants. The market is associated with high-risk investments, product differentiation, quality experience, and resilient marketing strategies. Since its commencement, GPRE has faced numerous setbacks including poor performances and marketing barriers due to international legal, economic, and political factors (Bargorett & Williams, 2014). Even though there are lucrative external environmental factors encouraging investments in renewable energy, the risk of new entrants is relatively lower due to these complications.

Threats of Rivals

The competitive landscape in renewable energy in the industry contains significant inconsistencies. This is attributed to the slow growth of the energy sector, despite there being favourable external conditions. Currently, only a handful number of companies are well established in the US (Pedersen & Sanderød, 2015). However, the international and regional platforms contain several well-established companies that are likely to compete with the GPRE. Albeit coal, nuclear, oil, and gas-powered energy resources are facing a substantial decline, they may still have significant competition to GPRE. Currently, GPRE produces about 45% of domestic biofuel products, while the rest of fragmented industry account for the rest (Becker, 2016). This indicates a considerable competitive advantage over other the rivals in the market. Therefore, rivalry threats in this case range from medium to moderate.

Threats of Bargaining Power of Buyers

The bargaining power attributed to biofuel demand is relatively marginal in the development of renewable energy market. Consumer power is eroded by the need to comply with strict federal and international obligations. This suggests that the consumer's threat is relatively low concerning GPRE due to federal and international regulations (Pedersen & Sanderød, 2015).

Threats of Bargaining power of Suppliers

The suppliers of bio-resources in GPRE's operations is significantly high due to the number of investments put into agribusiness infrastructures. The company exclusively relies on agriproducts to generate biofuels. The failure to comply with suppliers' demands may lead to substantial adverse effects to GPRE, indicates that the bargaining power of suppliers is very high (Nasir, Arshad, & Xiaorui, 2016).

Threats of Substitutes

GPRE primarily deals with the production of biofuels in the renewable energy sector. Alternative energy investments are increasingly becoming lucrative as consumers seek to invest in low-cost energy resources. Other companies in the industry are also exploring geothermal, solar energy, and wind energy sources, in which each organisation seeks to specialise in a specific niche (Nasir, Arshad, & Xiaorui, 2016). Although the alternative energy sector is still in its early stages of development, GPRE faces a high threat of substitution from other fields in alternative energy resources.

GPRE's Competitive Strengths

GPRE has implemented a series of tactical business strategies in the energy sector to remain competitive. Some of these strategies include effective risk management, optimizing market opportunities, and increased diversification in renewable energy societies. Others include acquisition and integration abilities, operations efficiency, robust partnerships, and vertical integration (Becker, 2016).

Conclusion

Apparently, external factors and competitive advantages have inspired GPRE transformation from relying on oil and gas resources to producing renewable energy products. Inferences from PESTLE and Porter's five model strongly indicates that Green Plains Inc. was motivated into investing in the energy industry due to uncertainties facing the oil and gas sector and lucrative opportunities characterising investments in renewable energy segment.

References

Bargorett, E. & Williams, B., 2014. Macro-environmental Analysis for the Alberta Midstream Oil and Gas Sector.

Becker, T., 2016. 2016 Annual Report - Green Plains Annual Report. http://greenplainsannualreport.com/wp-content/uploads/2017/06/Green-Plains-2016-Annual-Report.pdf

Edomah, N., Foulds, C. & Jones, A., 2016. The role of policy makers and institutions in the energy sector: The case of energy infrastructure governance in Nigeria. *Sustainability*, *8*(8), p.829.

Goss, E., Millewski, J., & Strain, S., 2015. The Costs and Benefits of Public Power in Nebraska: An Investigation of Electricity Rates, Taxes, and Competitiveness. https://www.platteinstitute.org/Library/DocLib/Platte-Power-Study-12-28-2015.pdf

Kehbila, A.G., 2012. A Strategic Business Analysis Model for EcoXergy Solutions Energy and Greenhouse Gas Consulting Services.

Lantz, E., 2016. Economic Development Benefits from Wind Power in Nebraska: A Report for the Nebraska Energy Office. *US Department of Energy Publications*, p.24.

Liserre, M., Sauter, T. & Hung, J.Y., 2015. Future energy systems: Integrating renewable energy sources into the smart power grid through industrial electronics. *IEEE industrial electronics magazine*, *4*(1), pp.18-37.

Mathiyazhagan, M. & Ganapathi, A., 2011. Factors affecting biodiesel production. *Research in plant Biology*, *1*(2).

Maxson, S., Rotering, J., Clark, A., Biscardini G, ... & Banno, T, 2017. 2017 Oil and Gas Trends Adjusting business. www.strategyand.pwc.com

Nasir, N., Arshad, S. & Xiaorui, S., 2016. Factors affecting alternative automotive fuel industry. http://www.diva-portal.org/smash/get/diva2:345809/FULLTEXT01.pdf

Pedersen, F. & Sanderød, E.M., 2015. A company analysis and valuation of the solar energy corporation Q-Cells SE. *Copenhagen Business School*, *200*, p.132.

Prpich, G., Darabkhani, H.G., Oakey, J. & Pollard, S., 2014. An investigation into future energy system risks: An industry perspective.

YOUR KNOWLEDGE HAS VALUE

- We will publish your bachelor's and master's thesis, essays and papers

- Your own eBook and book - sold worldwide in all relevant shops

- Earn money with each sale

Upload your text at www.GRIN.com
and publish for free